# Theory
## INTO
# Practice

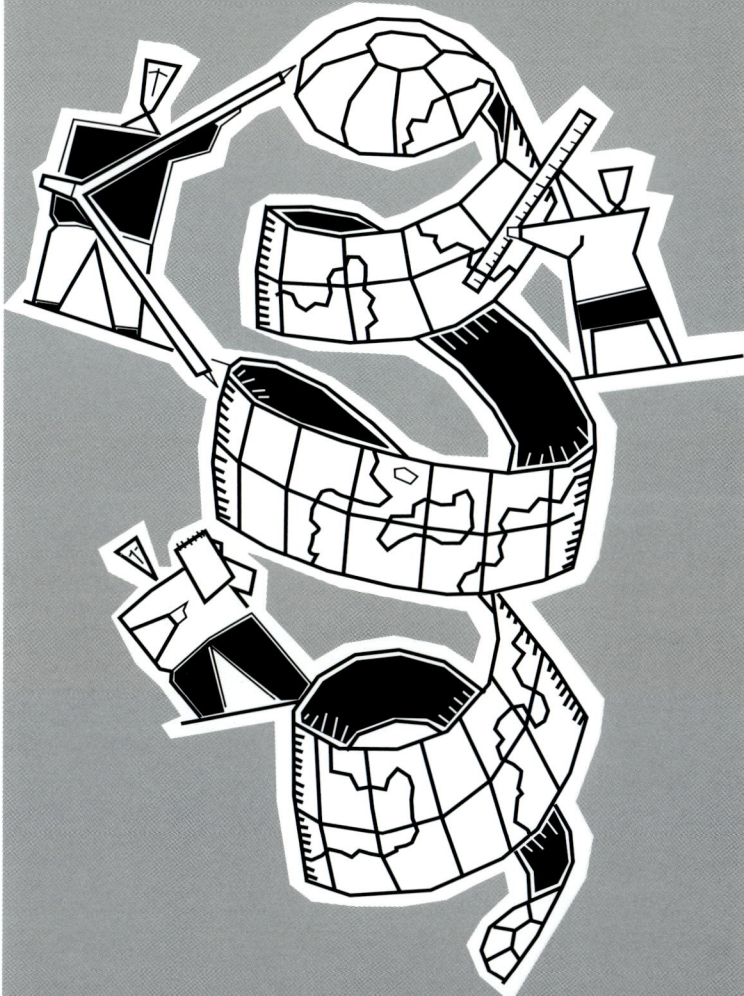

PROFESSIONAL DEVELOPMENT
FOR GEOGRAPHY TEACHERS
Series editors: Mary Biddulph and Graham Butt

# Moral
# Dilemmas

## MICHAEL McPARTLAND

Geographical
Association

## The author

Michael McPartland is Senior Lecturer in the School of Education, University of Durham.

## The series editors

Dr Mary Biddulph is Lecturer in Geography Education in the School of Education, University of Nottingham and Dr Graham Butt is Senior Lecturer in Geographical Education in the School of Education, University of Birmingham.

ISBN 1 899085 12 2
First published 2001
Impression number 10 9 8 7 6 5 4 3 2 1
Year 2004 2003 2002

Published by the Geographical Association, 160 Solly Street, Sheffield S1 4BF.
E-mail: ga@geography.org.uk
Website: www.geography.org.uk
The Geographical Association is a registered charity: no 313129.

The Publications Officer of the GA would be happy to hear from other potential authors who have ideas for geography books. You may contact the Officer via the GA at the address above. The views expressed in this publication are those of the author and do not necessarily represent those of the Geographical Association.

Designed by Ledgard Jepson Limited
Printed in Hong Kong through Colorcraft Ltd.

Geographical
Association

Photo: Michael McPartland.

# Contents

**Editors' preface**     **7**

**Introduction**     **8**

**1: What is a moral dilemma?**     **11**

**2: Do moral dilemmas really exist?**     **15**

**3: Moral dilemmas in geography**     **21**

**4: Embedding moral dilemmas in a narrative structure**     **25**

**5: Using moral dilemmas in geography**     **31**

**6: Creating a moral dilemma**     **37**

**7: Conclusion**     **41**

**Bibliography**     **43**

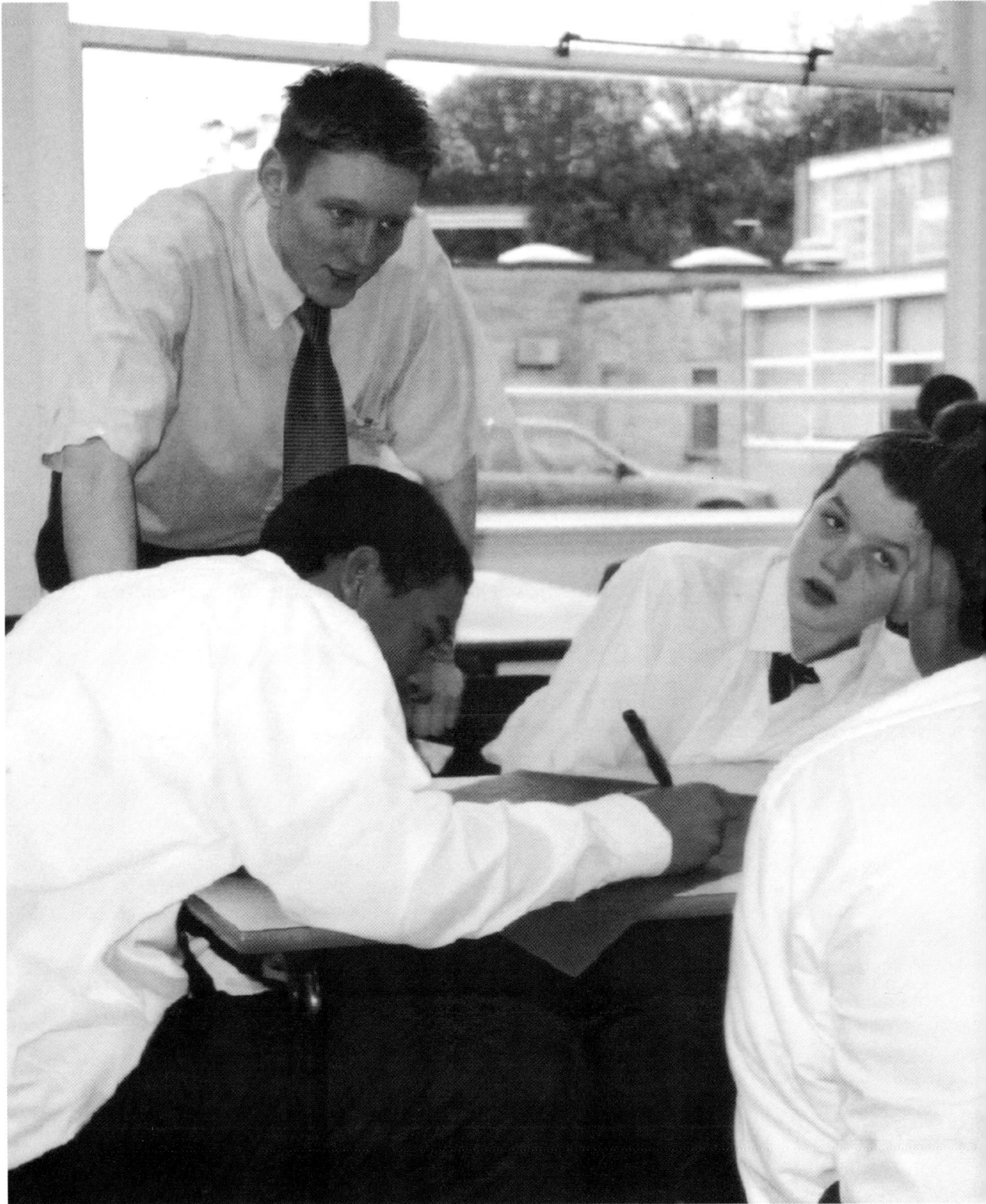

Photo: Michael McPartland.

# Editors' preface

*Theory into Practice* is dedicated to improving both teaching and learning in geography. The over-riding element in the series is direct communication with the classroom practitioner about current research in geographical education and how this relates to classroom practice. Geography teachers from across the professional spectrum will be able to access research findings on particular issues which they can then relate to their own particular context.

## How to use this series

This series also has a number of other concerns. First, we seek to achieve the further professional development of geography teachers and their departments. Second, each book is intended to support teachers' thinking about key aspects of teaching and learning in geography and encourage them to reconsider these in the light of research findings. Third, we hope to reinvigorate the debate about how to teach geography and to give teachers the support and encouragement to revisit essential questions, such as:

- Why am I teaching this topic?
- Why am I teaching it in this way?
- Is there a more enjoyable/challenging/interesting/successful way to teach this?
- What are the students learning?
- How are they learning?
- Why are they learning?

This list is by no means exhaustive and there are many other key questions which geography teachers can and should ask. However, the ideas discussed and issues raised in this series provide a framework for thinking about practice. Fourth, each book should offer teachers of geography a vehicle within which they can improve the quality of teaching and learning in their subject; and an opportunity to arm themselves with the new understandings about geography and geographical education. With this information teachers can challenge current assumptions about the nature of the subject in schools. The intended outcome is to support geography teachers in becoming part of the teaching and learning debate. Finally, the series aims to make classroom practitioners feel better informed about their own practice through consideration of, and reflection upon, the research into what they do best - teach geography.

**Mary Biddulph and Graham Butt**
**January 2001**

# Introduction

The interest, in England and Wales, in values education in general and in spiritual, moral, social and cultural values in particular is reflected in the flow of publications from government agencies concerned with these matters (NCC, 1993; Ofsted, 1994; QCA, 1997, 1998, 1999). In relation to the moral domain it is clear what these reports are advocating:

- The need to reassert the view that the education of young people must involve addressing their values and attitudes as a response to what is perceived to be a decline in their moral standards.

- The need to define a set of core values that can then be promoted in school both on a whole-school basis and within the context of different subjects. An attempt to do this is made in two Qualifications and Curriculum Authority (QCA) documents, *The Promotion of Students' Spiritual, Moral, Social and Cultural Development* and *Education for Citizenship and the Teaching of Democracy in Schools* (QCA, 1997, 1998). The former tries to identify and classify such core values under four headings, which relate to:

1. the self (e.g. develop self-respect and self-discipline),

2. relationships (e.g. resolve disputes peacefully),

3. society (e.g. promote opportunities for all), and

4. the environment (e.g. accept our responsibility to maintain a sustainable environment for future generations).

The second document lists a set of values and dispositions which might underpin the concept of citizenship, e.g. judging and acting by a moral code.

In a report entitled *Preparing Young People for Adult Life*, QCA (1999) offers the view that the importance of personal, social and health education in the curriculum is signified by the way it helps students to identify their values and live up to them. This report stresses the importance of getting young people engaged in the process of moral reasoning through the medium of *moral dilemmas*, which is the main focus of this book.

Geography teachers are engaging with the discussions that have followed the publication of the QCA documents and, in particular, responding to the question 'How can the subject make a significant contribution to the moral education of young people at the beginning of the twenty-first century?' According to Proctor and Smith (1999) these discussions are taking place at all levels of geographical education. As geography is impregnated with moral values, a central task for geography teachers is to focus more explicitly on them. If ethics can be defined as the systematic reflection on moral questions or specific moral concerns, within defined geographical contexts, then all geography teachers are engaged in an ethical endeavour. Indeed, Lickona has advocated building an ethical dimension into all subjects, noting that a 'value centred curriculum mainstreams moral education, moving it into the very centre of teaching and learning' (1991, p. 26).

This book is founded, therefore, on the following interlinked propositions:

1. That the moral dimension to many of the issues examined in geography ought to have a more explicit focus in the teaching of the subject. As Wright states: 'There are moral and spiritual themes everywhere - it is only we, the geography teachers, who stop them emerging' (2000, p. 41).

2. That moral dilemmas provide a useful device for responding to proposition 1 (above) because they combine both content (linked to specific geographical and environmental issues) and a process of reasoning deployed in an attempt to resolve them.

3. That for this device to be effective, the moral dilemma should be embedded within a narrative structure. This provides the richly detailed and realistic context, particularly if the narrative structure relates to people, places and the environment within which the moral dilemma is located.

4. That by focusing on moral dilemmas not only will we be inviting young people to identify the moral dimension in the geographical or environmental issue and to make and justify moral judgements in response to it, but we will also be involving them in the process of moral reasoning. Pring has noted that 'the ability to reason about and to resolve moral dilemmas, and the quality of that reasoning, are quite clearly important aspects of personal development' (Pring, 1984, p. 77). In a sense this book is arguing that thinking skills in geography (Leat and Nichols, 1999) should also embrace the concept of 'moral thinking' (Wilson, 1990).

5. That an enhanced capacity for moral reasoning will be transferred into other contexts both within geography and other curriculum areas.

I will explore these propositions by introducing two moral dilemmas: 'Bali Bound' and 'Pilgrim at Topanga Creek', both of which have been used in geography classrooms. The first, 'Bali Bound' (page 16), uses the theme of international tourism to illustrate the concept of a moral dilemma, and the second, 'Pilgrim at Topanga Creek' (pages 28-29 and 33), focuses on the topic of international migration. It illustrates the nature and quality of moral reasoning linked to the use of a moral dilemma in geography.

Source: J.W.Taylor, in *Thin Black Lines* (DEC Birmingham, 1988)

# 1: What is a Moral Dilemma?

The first step in understanding the role and value of moral dilemmas in geography teaching is to try and establish what they are.

The standard definition of a moral dilemma includes all those situations and only those situations where, at the same time, a person *ought* to adopt each of two alternatives separately but cannot adopt both together. The word 'ought' implies that there are reasons to support both courses of action or decisions and that this imposes moral requirements on the person to do so. It is the conflict that arises from the juxtaposition of these moral requirements, which lies at the heart of the moral dilemma (Sinnot-Armstrong, 1988). In essence, therefore, a moral dilemma exists in any situation where at the same time:

- there is a moral requirement for the person engaged in the dilemma to adopt each of two alternative courses of action or decision,

- neither moral requirement is over-ridden in any morally relevant way,

- the person cannot adopt both alternatives together, and

- the person can adopt each alternative separately.

I will apply this standard definition to a specific geographical example in order to illustrate the key features of a moral dilemma.

## A moral dilemma

On 14 November 1999 an article by Emma Cook appeared in the *Independent on Sunday* under the general theme of 'Ethical tourism'. An extract from this article is shown in Figure 1. In the previous month, *The Observer* (24 October 1999) produced its annual Human Rights Index (HRI) which identifies, using ten indicators, those countries which systematically abuse the human rights of its citizens. Figure 2 illustrates how the *Observer's* HRI is constructed and Figure 3 shows how selected nations were ranked according to this Index.

Amensty International is a worldwide movement of people who campaign for human rights. The organisations' appeals for victims of human rights violations are based on accurate research and on international law. Amnesty International is independent of any government, political ideology, economic interest or religion.

**Figure 1:** Extract from article on the ethics of visiting countries with a poor human rights record.

# Sun, sea and a little ethnic cleansing
## by Emma Cook

Travel agents are meeting next month to ask whether we should holiday in countries with poor human-rights records.

Where would you choose to spend an exotic holiday? Sunbathing on a tropical beach with endless white sands fringed with palm trees? Or maybe you'd appreciate something a little more cultural – a mountain trek and a tour of ancient ruins and remote villages? That's the sort of choice we're used to making.

But what if that paradise beach happens to be just miles from an entrenched civil war? Or an exotic island's business interests are controlled by a military regime?

Some of the most unpalatable regimes in the world are to be found in countries desperate to attract tourists and they are keener than ever to persuade us that they have some of the most desirable holiday destinations on offer. Take Burma for example. Next month it will be promoting itself as an ideal place for an exotic break when it takes a stand at the World Travel Market in London's Earl's Court. There will be no mention of its record of oppressing human rights and its lack of democracy. Burma will be in good company at the Market. There will also be displays by Indonesia and China.

But does it matter if we holiday in such countries?

**Figure 2:** *The Observer* Human Rights Index, 1999.

| Country | Population (millions) | Extra -judicial executions | Disappearances | Torture/ inhumane treatment | Deaths in custody | Prisoners of conscience | Unfair trials | Detention without charge/ trial | Executions | Sententence of death | Abuses by opposition groups | Total |
|---------|---------|---------|---------|---------|---------|---------|---------|---------|---------|---------|---------|---------|
| Congo | 49.2 | 0 | 3 | 3 | 0 | 3 | 3 | 3 | 3 | 3 | 3 | 24 |
| Rwanda | 6.5 | 3 | 3 | 3 | 2 | 0 | 2 | 3 | 2 | 3 | 3 | 24 |
| Burundi | 6.6 | 3 | 2 | 3 | 3 | 0 | 3 | 3 | 0 | 3 | 3 | 23 |
| Algeria | 30.2 | 3 | 2 | 3 | 0 | 2 | 3 | 3 | 0 | 3 | 3 | 22 |
| Sierra Leone | 4.6 | 3 | 0 | 3 | 2 | 1 | 2 | 3 | 3 | 3 | 3 | 22 |

**Figure 3:** How some nations rate on *The Observer* Human Rights Index.

| Rank | Country | Population (millions) | *Observer* Human Rights Index |
|------|---------|---------|---------|
| 1 | Congo | 49.20 | 24.0 |
| 6 | Egypt | 65.70 | 21.0 |
| 9 | Indonesia | 206.50 | 20.5 |
| 46 | Sri Lanka | 18.50 | 11.0 |
| 77 | Greece | 10.60 | 6.5 |
| 88 | Australia | 18.40 | 5.0 |
| 126 | UK | 58.20 | 3.5 |

These materials can be used as background information when we ask students to respond to the following moral dilemma:

'You live in Indonesia and have just graduated from university with a good degree in English.

Your father is a priest of the local temple and an active member of the local *banjar* (village community). He dislikes the invasion of tourists who come to see important temple festivals.

You have found it difficult to find a job in Bali, but would like to stay. The only job is in the government tourist office.

Should you work for them?'

Source: Mason, 1995, p. 72.

# Relating this moral dilemma to the standard definition

How does this moral dilemma (however imperfectly stated in terms of the level of contextual detail) relate to the standard definition provided above? There are a number of factors that should help us clarify the arguments as to how, in deciding either to accept or refuse the job, the graduate (male) has to balance the moral requirements that underpin his decision.

- He has to earn money so that he can provide for his own economic needs and those of his family, including his father, as well as contribute to the local economy. This is especially true in a situation where there are few job opportunities. By taking this job he will assist the Bali tourist industry to grow and become more efficient and render a more effective service for its visitors from overseas. Perhaps he feels there is a moral requirement to take the job in order to exploit the advantages of possessing the university degree for which he has worked so hard.

- He must balance these moral considerations against the need to respect the social role which his father performs allied to his father's religious beliefs linked to that role. His father believes that tourism is undermining the cultural norms of Balinese society, leading to a distortion of its values and he does not want his son to be involved in the tourist industry.

This example indicates that although there are moral reasons and therefore moral requirements linked to both courses of action, these moral requirements do not over-ride each other in any significant or morally relevant way. The son can adopt either course of action, because there are no apparent impediments to this, but he cannot adopt both at the same time.

The son is faced, therefore, with what appears to be a moral dilemma according to the standard definition advanced on page 11. However, there are those who dispute the possibility that such a dilemma could exist; which is the focus of the next chapter.

# 2: Do moral dilemmas really exist?

Let us try to examine the question of whether moral dilemmas really exist by looking at a 'real' moral dilemma: 'Bali Bound' (Figure 4). 'Bali Bound' relates to the theme of the impact of tourism on less developed economies – one already used to illustrate a standard definition of a moral dilemma (page 11-13). This has been constructed in order to illuminate the nature of the debate and incorporates information provided on page 11.

## Questioning the existence of moral dilemmas

Among moral philosophers there is dispute as to whether moral dilemmas actually exist. Gowans (1987) argues that the proponents and opponents of moral dilemmas can be distinguished by the styles of reflection which they bring to bear on any moral situation. He makes a distinction between two styles of moral reflection: rationalist and experientialist. Those who reject the existence of moral dilemmas are called rationalists; those who accept the existence of moral dilemmas are experientialists. The 'Bali Bound' moral dilemma is used to illuminate the nature of these two styles of moral reflection.

### Rationalists

Rationalists regard the practice of moral decision making as being subject to the dictates of human rationality. In the western tradition of philosophy, rationalism is distinguished by the application of abstract principles and concepts to any moral situation. These then act as a clear guide for moral decision-making. The rationalist regards the particulars of the context of the moral experience as a source of distraction; rather he or she seeks to identify its generic features, since the application of principles and concepts to these features would help an individual to make a clear, consistent and defensible moral judgement. Any apparent moral dilemma merely reflects some inconsistency in the application of principles linked to a coherent moral theory or code – a theory or code with universal applicability. In relation to 'Bali Bound' a rationalist would argue that the principle of respecting the religious and cultural diversity of a society is of prime importance. This principle is apparently being challenged in Indonesia. If it were rationally applied to the 'Bali Bound' dilemma it would lead Michael and Li Shan to one inevitable conclusion: they should refuse to go on holiday to Indonesia.

# Bali Bound

My name is Michael Drumlin. I am a novice teacher teaching geography at a secondary school in Durham City. My wife is called Li Shan. She is Chinese and was born in Hong Kong 22 years ago. Her parents emigrated from Hong Kong just before it was reunited with China. Her mother is an engineer and her father has a small import-export business dealing mainly with companies in South East Asia. He still has strong links with the Chinese communities in the region.

I studied geography at university. I met my wife there. We met at an Amnesty International fund-raising event since we are both concerned about the lack of basic human rights in many parts of the world, including our own.

My wife studied ecology at university. She works as a research assistant for the local Water Company. I do not have much spare time. I spend a lot of time preparing lessons, marking books and developing new resources for my teaching. I really need and enjoy my holidays because by then I am exhausted. I love travelling and listening to music.

Last Monday evening I arrived home about 5 o'clock and found to my astonishment that I had won a competition – my first ever! Three months ago I entered a competition in the newspaper. It involved writing a tourist slogan for a well-known tourist company who wanted to encourage people to choose the island of Bali in Indonesia as a holiday destination. I entered this competition with the slogan 'I am Bali Bound for a beautiful holiday!' because I have a special interest in South East Asia. I made a study of it at university. I even wrote a final-year essay on the impact of tourism on Bali.

My wife and I could not believe that we had won this one-week holiday on the island, staying at the Le Meridien Nirwana Resort on the south-west coast. The brochure which we received from the company boasts views of Agung Volcano, the rice terraces, sunsets over the Indian Ocean and a nearby 18-hole championship golf course designed by Greg Norman. We could also stay an extra week if we wanted, for half the price of a normal one-week holiday in the hotel. We planned to take the holiday during the coming Easter break.

Soon after we received this great news I decided to use the Internet to get more information on Bali and Indonesia. I was shocked to discover a Human Rights Watch report. In this report it described the way in which the Chinese minority in Indonesia, mostly small shopkeepers, was being attacked and injured. Prices have risen sharply in recent years and the Chinese community is being blamed for the economic problems of the country. We were particularly upset at a report which stated that Chinese women had been raped during the May riots last year. I subsequently learned that, in 1999, Indonesia was ranked ninth out of a list of 194 countries for its abuse of human rights. Further investigations revealed, much to my wife's dismay, that tourists were damaging the coral reefs and that sewage from the newly built hotels was flowing back on to the beaches.

My wife, her family and I are deeply concerned and we do not know if we should go ahead with the holiday to Bali. What ought we to do?

UNIVERSITY OF HERTFORDSHIRE LRC

### Experientialists

Experientialists try to understand moral dilemmas and conflicts from the standpoint of the moral experience of the individuals involved in the process. Experientialists take account of the importance of the social context within which the moral decision has to be made, to the values and attitudes of the participants involved, to the complexity of the situation as well as the consequences of the final decision. Unlike rationalism, experientialism stresses the particulars of the moral context, the crucial role of emotion in decision-making and the importance of observation and reflection in any moral situation. In relation to 'Bali Bound', an experientialist accepts the moral complexity of the situation and would defend Michael and Li-Shan's right to evaluate the moral requirements linked to each course of action and the decision made.

Within the context of experientialism a number of arguments have been advanced to justify the claim that moral dilemmas do exist. One such argument (presented below) is related to the existence of a plurality of values.

# Plurality of values

This argument is constructed on the premise that there is a plurality of moral values and that it is inevitable, given the world in which we live and the situations we encounter, that these values will sometimes come into conflict. To illustrate this argument we can apply the values formulated in *The Promotion of Students' Spiritual, Moral, Social and Cultural Development* (QCA, 1997) to the dilemma considered by Michael Drumlin and Li Shan. This QCA publication resulted from a recommendation made by the National Forum for Values in Education and the Community. The NFVEC recommended that efforts should be made to try and identify any values upon which there might be common agreement in society. These were designed to support schools in the important task of contributing to the spiritual, moral, social and cultural development of their students. The values were structured around four themes: 'the self', 'relationships', 'society' and 'the environment'; and have been included in the handbooks for secondary and primary teachers in England (DfEE, 1999a,b). We can draw upon this list in order to help Michael and Li Shan decide, with respect to their proposed holiday to Bali, whether they should either:

1. as professionals, 'strive throughout life for knowledge, wisdom and understanding' and take the holiday? *(self)* or

2. 'care for others and exercise goodwill in our dealings with them' – including, presumably, their families – and stay at home? *(relationships)* or

3. 'refuse to support [by going to Bali] values or actions that may be harmful to individuals or communities – including the Chinese community in Indonesia – and stay at home as an act of solidarity? *(society)* or

4. 'contribute to, as well as benefit from, economic and cultural resources' and go on the holiday because of the contribution it might make to the Indonesian economy? *(society)* or

5. stay at home because, conscious that tourism is increasingly having a negative impact on the environment, they wish to 'preserve balance and diversity in nature wherever possible' *(environment)*.

Source: DfEE, 1999a, pp. 196-7.

*Opposite page*
**Figure 4:** 'Bali Bound' a moral dilemma relating to the impact of tourism.

When we try to apply these values to Michael and Li Shan's situation it becomes obvious, given that the values identified may be called upon both to support and oppose the holiday, that they are in a situation of moral conflict.

The essence of this conflict or dilemma can be presented diagrammatically. In Figure 5 the rights and duties linked to the principles located in each of the four domains identified by the QCA (self, relationships, society and environment) are juxtaposed.

| | Self | Relationships | Society | Environment |
|---|---|---|---|---|
| Self | | | | |
| Relationships | Michael's right to a holiday versus the need to respect his and Li Shan's families anxieties | | The need to highlight the dangers of inappropriate tourism against the need to promote cross cultural relationships | |
| Society | Michael's right to become a more knowledgeable teachers against the need to show solidarity with the Chinese community in Indonesia | | | |
| Environment | Michael and Li Shan's right to hygienic conditions in the hotel versus the need to protect the coastal flora and fauna from pollution | | The need for the Indonesian people to develop their tourist economy against the need to protect the coral reefs | |

**Figure 5:** 'Bali Bound' Michael Drumlin and Li Shan's dilemma related to the self, relationships, society and the environment.

The prior discussion might lead one to the conclusion that an experientialist style of moral reflection is a more defensible position to adopt in that it acknowledges:

- the importance of the context of the situation,

- the lack of full knowledge of the circumstances of the situation held by those involved in the dilemma,

- the inevitable emotional dimension to the decision to be made, and

- the intrusion of values and attitudes into the decision making process.

One response to the 'Bali Bound' dilemma might be to try and distinguish between moral and non-moral values, arguing that though there may be different and conflicting non-moral values, there must always be one pre-eminent moral value and that this would take precedence over all non-moral values. At issue here, and permeating much of the debate about moral dilemmas, is the larger concern about what is the difference between moral and non-moral values. What makes any environmental and geographical issue a more issue? This question is the focus of the next chapter.

Photo: © Greenpeace UK.

# 3: Moral dilemmas in geography

In an article in *The Guardian* (26 April 2000) headlined the 'Trials of an unlikely eco-warrior', Lord Melchett (then Executive Director of Greenpeace), in a discussion about genetically modified foods claims that:

> *'Greenpeace doesn't have a moral objection to genetically modified foods. We don't believe it is something that religion wouldn't countenance, or that it is an interference with nature to which people would have an ethical objection. We just think there is a danger in introducing something into the environment that you can't recall'.*

Notwithstanding the provocative way in which Lord Melchett equates morality with religion, and the assertion that any interference with the natural order is not in itself an ethical matter, his statement does lead us to reflect on the following questions.

1.  What makes an issue a moral one?

2.  Are dilemmas embedded within an issue moral dilemmas?

3.  What characteristics might a moral issue or dilemma embody?

If we relate these questions to 'Bali Bound', 'What is it about the situation that Michael and Li Shan find themselves in, that makes it a moral dilemma?'. We can try to establish the grounds on which we might claim that Michael and Li Shan's decision to take or not to take a holiday in the circumstances described on page 16 is a moral one. Some might argue that the question is not worth pursuing since the concept of morality is too ambiguous, imprecise, elusive or contestable. However, if one of the central purposes of introducing moral dilemmas into geography teaching is to encourage students to identify the moral dimension to a geographical or environmental issue, then this presumes that both we, as geography teachers, and our students have some understanding of what 'moral' means, however difficult it may be to arrive at that understanding. We can try to do this by structuring the discussion around a number of propositions.

## Proposition 1

The moral requirements linked to each course of action or decision in a moral dilemma are underpinned by *values* which are moral because they can be applied to more contexts than the one in which they are presently being applied. In essence they are *universal*. Thus, for example, the need to show respect and concern for the beliefs, feelings and experiences of other human beings and to *empathise* with their concerns must be considered a universal requirement. In this case we must show respect and concern for the immediate families of Michael and Li Shan, who are anxious about the *consequences* of their decision to take a holiday. A moral reason goes beyond acting in self-interest – Michael might wish to claim that as a professional committed to upholding standards as a geography teacher he has an *obligation* to visit Bali as a means of widening his geographical horizons as well as deepening his geographical knowledge.

## Proposition 2

If we accept the notion that the moral requirements and the values which underpin them can be universalised then we can claim that these requirements *ought* to be applied in other similar contexts and that by doing or failing to do this allows us to apply concepts of *right or wrong* to the decision. If we accept, as a moral requirement, the need to respect the *human rights* of all people then Michael and Li Shan's decision not to holiday in a country where such rights are being abused (Indonesia) becomes both a moral *duty* and a moral imperative.

## Proposition 3

In the process of incorporating this principle into their moral vision, we would expect Michael and Li Shan to act in accordance with that vision. In other words, to act in accordance with their *conscience* when a similar situation arises in the future, leading to a level of consistency in their *behaviour*. We are also assuming that Michael and Li Shan would be able to offer *moral reasons* to justify their decision, so that their decision becomes, in the true sense of the word, reasonable.

These three propositions embody certain concepts: *right or wrong, consequences, universality, conscience, value, principle, ought, obligation, empathy, human rights* and *behaviour*. If, during a whole-class discussion about a specific geographical or environmental issue, we invoke this assembly of concepts then we might claim that the reasons advanced in the discussion (to justify a point of view or course of action) become moral reasons and the requirement to act in specific ways becomes a moral requirement. As it is not difficult to apply these concepts to the dilemma faced by Michael and Li Shan; theirs is, therefore, a moral dilemma.

Photo: © Associated Press AP.

# 4: Embedding moral dilemmas in a narrative structure

The central assertion in this chapter is that moral dilemmas are best considered when embodied in the context of a narrative structure. Narrative is the most personal mode of discourse. It involves a person recounting a story to a reader or listener. This person is the central participant in the story and describes the events from the perspective of an observer (first or third person). The statements are arranged chronologically since they relate to past events and are specifically oriented to the actors in the narrative (Longacre, 1992).

Narrative, as a mode of discourse, has spawned a range of genres to meet the needs of different audiences and to serve different social purposes: myths and fables, eye-witness accounts, legends and fairy tales, narrative poems and ballads, parables and mysteries, biographical sketches, anecdotes, drama, letters, and novels which all reflect, in part, a narrative structure.

This chapter demonstrates how narrative may be used to explore the moral dimension to a geographical theme and the moral dilemmas located within the theme. It takes as the start point a unit of work on migration, examined in a popular textbook resource used for year 8 students in geography (Waugh and Bushell, 1992).

## Starting from a geographical theme

One chapter of the geography textbook *Connections* (Waugh and Bushell, 1992) focuses on Population issues, and contains a sub-section on the theme of migration. The concept is defined, a simple typology of migration is presented and the causes of migration are explained (using the push and pull framework). The consequences of international migration for Mexico and the USA are examined as a case study. The textbook uses three kinds of discourse to explore the topic of migration:

- *expository discourse* is used to explain concepts such as rural to urban migration, for example, with accompanying factual detail,

- *argumentative discourse* - allows participants in the migration process, or those affected by it, to present their case as to why it has been advantageous or disadvantageous to be involved in the process or have it impinge on their lives, and

- *procedural discourse* is used to instruct the students, for example, on how to draw a bar graph to illustrate the growth of Mexico City.

I would argue that the process of migration and its consequences are imbued with moral concerns. At a general level the migration flow between Mexico and the USA (both legal and illegal) is a moral issue since it springs from deep inequalities in human welfare, a result of disparities in resource availability, production and consumption, in economic opportunities, in social amenities and political power for which migration may be regarded as a 'corrective' human response. In addition many of the poor migrants to the USA are not valued, are isolated, lack equal opportunities, and benefit little from the economic and cultural resources available. They find it difficult to make use of their talents, are frequently denied the basic human rights of adequate shelter, education and health and are forced to live a less than dignified lifestyle. It is not difficult to attach a whole range of moral concepts to this process and it is this which invests the flow of migrants from Mexico to the USA with a moral dimension.

Despite outlining the facts *Connections* does not explicitly invite the teacher to focus on migration as a moral concern. Of course this was probably never the prime intention of the authors of the textbook, in which case the responsibility of the geography teacher to do so assumes some importance. However, to give the moral dimension a more explicit focus, a moral dilemma on the theme of migration should be integrated into the unit of work, and it needs to be embedded in a narrative structure to supplement the modes of discourse (expository, argumentative and procedural) used in the textbook.

# Narrative discourse

We can illustrate the value of using narrative as a vehicle for exploring moral issues by adapting material from a novel. *Tortilla Curtain* (Boyle, 1995) vividly captures the social and economic divide between two worlds: the world of the wealthy, white Americans of the Los Angeles region - represented by Delaney Mossbacher and his family - and the world of the poor, Mexican migrant - represented by Candido Rincon and his wife America. The moral dilemma constructed using the opening event from the novel *Tortilla Curtain* is shown in Figure 6. It has been entitled 'Pilgrim at Topanga Creek'.

How can we justify the use of this narrative extract to supplement information from the textbook *Connections* in the geography classroom? Why is narrative such a powerful medium for examining the moral dilemmas that are embedded in many geographical issues and the complex moral decision making which goes with it? There are, perhaps, three reasons to justify this.

In an explicit way the fundamental nature of narrative and the values which it attempts to transmit are embodied in its structure. Labov (1972) has examined the formal structure of narrative and identified a number of crucial elements, including:

1. the orientation element, which provides information relating to the context of the narrative – the time, place and persons involved in the narrative – as well as the general background information,

2. the complication element, which identifies the conflict at the heart of the narrative, and

3. the evaluation element.

For Labov the 'evaluation element' is the means by which the narrator indicates the point or moral of the story. The events of the Topanga Creek moral dilemma have been configured deliberately to force the narrator, Delaney, to make a moral decision that he subsequently attempts to justify to himself. White claims that:

*'every ... story ... is a kind of allegory, points to a moral or endows events ... with significance ... every historical narrative has, as its latent or manifest purpose, the desire to moralise the events of which it treats'* (1981, pp. 13-14).

The event occurring at a particular time and in a specific place – in this case on the edge of the Topanga Canyon – would not have occurred if deep spatial inequalities in social and economic welfare had not prompted it – in this case the migration of the Mexican to Los Angeles.

Tappan has argued (1990) that all moral experiences and the moral decision making associated with them incorporate three interdependent psychological dimensions: the cognitive, the affective and the conative - the process of thinking, feeling and willing. In any moral dilemma the central agent, when faced with the challenge to make and justify a decision, is influenced by these three inter-related dimensions. Delaney Mossbacher's decision is influenced by the prior knowledge and prior perceptions which he brings to the decision, by an emotional stance which heavily informs his response and by a predisposition to act in one way or another shaped by his knowledge, understanding and feelings. Bruner has suggested (1986) that one of the prime functions of narrative is to hold cognition, emotion and action together and is a particularly powerful mode of discourse for doing so.

Narrative is all-pervasive in our lives. For many it is the mode of discourse most natural to us (Hardy, 1968). When challenged to justify a decision we have made, including a moral decision, we do so with reference to the narrative context within which that the decision was located. For McIntyre (1981) any action only becomes morally intelligible when we locate its place within a narrative. Thus, in inviting students to make and justify a decision which they think a person ought to take, we, as teachers, are not imposing on them an alien form of discourse. Rather narrative mirrors closely the kind of process in which we all indulge when attempting to justify our own moral decisions. Narrative is a powerful mode of discourse for discussing moral dilemmas because it captures the inherent complexity of moral decision making.

# The Pilgrim at Topanga Creek
**Delaney's story**

My name is Delaney Mossbacher. I live at 32 Pinon Drive, Arroyo Blanco estate, a few miles west of Los Angeles close to the Santa Monica Mountains. I live there with my second wife Kyra, our son Jordan, our two terriers and a Siamese cat. The estate is located near the side of a ridge which overlooks a deep valley called the Topanga Canyon. It is a private estate. It has its own golf course, ten tennis courts, a community centre and 250 houses. All the houses are built in the Spanish Mission style. They are all painted in one of three shades of white and have orange roofs. Nobody is allowed to paint his or her house in a different shade.

Photo: Kathy Vilim.

I got up at seven that morning, as usual, to make Kyra's coffee, fed Jordan his fruit high-fibre bar and let the dogs out into the garden. It was a beautiful sunny morning. The temperature was about 30°C. I squeezed three oranges and, while Kyra sipped her coffee and washed down her 12 vitamin tablets with the juice, I made a cup of herbal tea and two slices of wheat toast for myself.

I am a journalist. I write a monthly column for a magazine called *Wide Open Spaces*. I am an environmental journalist. I write about changes to the wild flowers and animals of the region, day by day, season by season. The column is called 'Pilgrim at Topanga Creek'. I dedicate it to my late Aunt Dilliard. She was also an environmentalist and like me she loved nature more than she loved people. In my column I often express my concern for the pupfish, the Florida manatee and the spotted owl because they may all disappear soon. I worry a lot about over-population, deforestation, global warming and the 5 billion people using up the limited resources of our planet. This morning I heard on the radio that there are only 75 Californian condors left on earth.

After Kyra had driven Jordan to school in her car, I decided to visit the recycling plant at the top of the canyon. So I loaded my new Japanese car, just washed and waxed, with a pile of newspapers and empty diet-coke cans and drove out of the estate. The road winds up the canyon. It was about 10 o'clock. I was driving along when suddenly a man appeared in front of my car. I felt a bump and knew at once that I had hit him.

I am ashamed to admit that my first thought was for my car. Was it dented? Would my car insurance rate go up? And then I remembered him. Who was he? Where was he? Was he badly hurt? Was he dead? I thought he must have been hiding in the bushes at the side of the road and had decided to jump in front of the car. Even now I remember the look of fear on his face, the flash of his moustache and his cry.

I was trembling as I turned the engine off. I got out of the car in a daze. The dust was still billowing in the air because I had braked so quickly. Other cars passed by but none stopped to help. Perhaps they thought it was a set up. There had been reports of gangs of Mexicans faking accidents and then attacking and robbing the drivers as they got out of their car. To the left across the road was the wall of the canyon; to the right the canyon fell away to the dry sandstone bed of Topanga creek many metres below. I couldn't see anything other than sagebrush and treetops but I guessed the man was down there among the scrub oak and the manzanita bushes. I remember thinking why did this have to happen to me? What had I done to deserve this?

And then I heard a low moan. I looked into the bushes at the side of the road and there he was lying on the ground with blood coming from his mouth. One side of his face was badly bruised. He was still holding a plastic bag with some tortillas in it. He said something in a foreign language. I suddenly realised he was speaking Spanish and that he was probably an illegal immigrant from Mexico. He was probably living at the bottom of the canyon. Was he one of the Mexicans, I wondered, who mowed the lawns of the estate? I have lived in Los Angeles for two years but this was the closest I have come to a Mexican. Where had he come from? What did he want? Why had he thrown himself under my car?

I asked him slowly if I could help him. And then he smiled or tried to. A film of blood clung to his jagged teeth, half hidden by his moustache. He licked the blood away with his tongue. I asked him what he wanted. And then he whispered 'Money, money' rubbing the fingers of his undamaged hand.

**Question:** Ought Delaney to give him some money?

Photo: Michael McPartland.

# 5: Using moral dilemmas in geography

This chapter focuses on how moral dilemmas can be used in the geography classroom and looks at the nature and quality of the moral reasoning linked to the use of 'Pilgrim at Topanga Creek'.

The use of moral dilemmas for promoting moral reasoning in the geography classroom has been advocated as one element in a whole repertoire of teaching approaches linked to values education (Maye, 1984). Slater (1982) has offered advice on how such dilemmas might be used in the classroom as part of a range of approaches for the interpretation and analysis of values in geography and, in a more recent paper, has demonstrated effectively the multiplicity of ways in which geography is permeated with values (Slater, 1996). There has, however, been limited empirical research on measuring the efficacy of using moral dilemmas to promote moral reasoning in a subject-based curriculum (Blatt and Kohlberg, 1975; DeHaan *et al.*, 1997). Two conclusions which emerge from this research appear to indicate that promotion of moral reasoning is most effective when it is integrated into subject-based teaching and when the strategies employed by the subject teacher encourage structured discussion in the classroom.

The moral dilemma 'Pilgrim at Topanga Creek' (Figure 6, pages 28-29 and Figure 7, page 33) was used with year 8 students studying migration within the context of a case study of migration from Mexico to the USA. The students had already discussed the concept of migration, types of migration, the consequences of international migration on the host and recipient countries and the importance of the push and pull framework for examining the causes of migration. The moral dilemma was introduced using a phased approach.

## Phase 1: Presenting the dilemma

Copies of 'The Pilgrim at Topanga Creek: Delaney's story' (Figure 6) were handed out to the students. The teacher read the story to them and used an atlas and photographs to confirm the location of Topanga Canyon in relation to Los Angeles and to clarify specific concepts, such as 'canyon', 'creek', 'Spanish mission-style architecture', 'sagebrush'.

Each student was asked to consider the story and its attendant details carefully and to decide, as individuals, which decision (to give or not to give money) Delaney ought to make. The students recorded their decision and were asked to offer reasons to support their view.

# Phase 2: Analysing the dilemma

A teacher-led discussion on the alternative decisions made and the nature of the reasons advanced then took place. During this discussion students were encouraged to justify the alternative choices they had made, specific aspects of the story were clarified and, for the students, the complexity of the issue was explained.

Groups of three or four students, who had made a similar decision, were then asked to choose and justify the best reason for Delaney giving or not giving the Mexican some money, and to share their views with the rest of the class.

# Phase 3: Extending the dilemma

The moral dilemma 'Pilgrim at Topanga Creek' has two parts: the first part (Figure 6) tells the story from the American, Delaney's, perspective - culminating in the request for money by the Mexican. The second part (Figure 7) tells the story from the perspective of the Mexican, Candido - culminating again in his request for money.

Copies of Candido's story were handed out and read by the students. The students were then asked (individually) if, in the light of Candido's story, they had changed their mind as to whether Delaney should or should not give Candido money. They were, once again, encouraged to offer reasons to justify their decision. This information was recorded and debated during a teacher-led whole-class discussion.

# Phase 4: Reflecting on the contexts

Finally, the whole-class discussed the economic, historical and political contexts in which the dilemma was embedded. The students and teacher considered the ways in which these contexts might be changed to eliminate the dilemma.

# Investigating the moral reasoning

We can immediately ask:

- What was the nature and quality of the moral reasoning which accompanied the use of the first part of the dilemma in the classroom?

- How might we classify and interpret the nature of the response to the dilemma?

It is necessary to use some kind of framework for the analysis and implementation of the responses made. Wilson (1990) evaluates the nature and quality of the moral reasons advanced in response to moral conflict according to certain attributes or components, on the grounds that during a classroom activity such as the one described above, we are not looking for the 'correct' response but for the quality of the student's moral reasoning – i.e. the extent to which the student is being morally reasonable in the decisions he or she makes. Figure 8 lists the main components of Wilson's framework and uses a simple code to indicate what each one means.

*Opposite page*
**Figure 7:** Extending 'The Pilgrim at Topanga Creek' moral dilemma: Candido's story.

# The Pilgrim at Topanga Creek
## Candido's story

My name is Candido Rincon. I was born in Tepoztlan in the south of Mexico 33 years ago. I have a girlfriend called America. She is 17 years old and she is expecting our first child in four months time. She was born in a small village not far from Tepoztlan. I have known her since she was four years old. She is the youngest sister of my wife, Resurreccion. She was a flower girl at our wedding.

A few years ago I spent some time working in the potato fields of Idaho. In nine months I made more money than my father had made in his leather shop all his life. Most of the men in my village had gone north to the USA to work in the fields, tired of sitting all day in the cantina drinking beer. A few men stayed behind – the rich and the crazy and the men who stole your wife while you were away in the north. This is what happened to me. I came home to find that my wife had left to go and live in Cuernavaca with a man called Teofilo. She was six months pregnant and she had spent all the money I had sent her. America was the one who broke the news to me. I was so ashamed. I wandered the hills of the Sierra Juarez sleeping in my clothes. I tried to cross the border but the US immigration caught me and put me in prison in Tijuana.

When I came out I had no money. I danced for people in the street. I begged from the turistas. I stole a can of kerosene and became a tragafuegos – a streetcorner firebreather – earning a few centavos. One day I met America in the street. She was sixteen and she looked just like her sister only much better. I told her I was going to take her with me when I go north again. One month later we crossed the border at night and made our way to Los Angeles – the City of the Angels.

We had no money to rent a place to live. The streets of the city are dangerous and so we decided to live at the bottom of a canyon. We had a sandy bank near the river to live on, wood for a fire, a stream for drinking water and for washing and when it rained a tarpaulin sheet hung between the trees kept us dry. America was worried about the snakes and the spiders. Each day I left the canyon to look for work. It was not easy but at least in the canyon we were safe from la chota – the police – and the immigration. I managed to find some work building walls or clearing sagebrush from ravines. We had very little money.

One morning in August I decided to visit the small shop close to the canyon to buy some tortillas. America had built a small fire. The hot sun climbed above the canyon walls. America said she would make some tea from the manzanita berries. We always boiled the water before drinking it since the rainwater had drained through the septic tanks of the large houses of the area before entering the creek. I made my way up through the canyon to the road. I kept my eyes down, not wanting to look at the gringos on their way to work in their cars. To them I was invisible. I reached the Chinese grocery after about half an hour and bought a stack of tortillas to go with the pinto beans.

I returned along the road thinking of America. Was she alright deep in the canyon among the chapparal and the oak trees? At first I did not know what hit me. I was flung into the bushes at the side of the road. I was in great pain. My face seemed on fire and my arm was hurting badly. I soon realised I had been hit by a car – a gringo's car. I saw him standing over me. He asked me if he could help me. I knew this was my big chance and I smiled with happiness. 'Money', I said, 'give me money'.

| Code (see below) | Main component |
|---|---|
| Phil (HC) | Having the concept of a person |
| Phil (CC) | Claiming to use this concept as a universal, over-riding and prescriptive moral principle |
| Phil (RSF) (DO) and (PO) | Having feelings which support this principle either of a 'duty oriented' (DO) or 'person oriented' (PO) kind |
| Emp (HC) | Having the concept of various emotions |
| Emp (C) and (UC) | Being able in practice to identify one's own emotions and other people's both at the conscious and unconscious levels |
| Gig (KF) and (KS) | Knowing the relevant facts and sources of facts |
| Gig (VC) and (NVC) | 'Knowing how': a skill element in dealing with moral situations both in terms of verbal and non-verbal communication |
| Krat | An alertness to moral situations and to make moral judgements and implement them |

**Phil** is an attitude which regards others as equals, accepting their interests as equally important.

**Emp** is an ability to know what oneself and others are feeling and what their interests are.

**Gig** relates to the attainment of knowledge of the facts relevant to any moral choice and an ability to know how to perform effectively in any social context.

**Krat** relates to the motivation to draw upon these other moral components and act upon a moral judgement made.

**Figure 8:** Wilson's moral components.

The student's responses to the 'Pilgrim at Topanga Creek' activity were classified according to the degree to which they exemplified Wilson's components. The degree to which they do so is a measure, according to Wilson (1990), of the quality of that student's moral awareness and moral thinking. This framework can also be used to identify both a hypothetical response for each of the categories and, in most cases, an actual response to the 'Pilgrim at Topeka Creek' dilemma presented to the year 8 students (Figure 9).

Wilson's list of moral components has been criticised because it emphasises the form of reasoning but neglects the context within which that reasoning takes place (Bottery, 1990). I would argue that the value of using moral dilemmas in geography in a narrative context, rich in detail, is that it links precisely the form of reasoning to that context.

In terms of the quality of moral reasoning, a high-quality response would embody some of the following attributes, demonstrating that the student:

- possesses some idea of what constitutes a person and accepted that certain rights flowed from that recognition;
- understands that these human rights must apply in all similar circumstances;
- is aware that such feelings as anger, remorse, pleasure, disapproval or sorrow are linked to the acceptance or rejection of these human rights;
- has an appreciation of the ways his or her own emotions and those of others are inextricably bound up with the situation;
- is using judgements that are informed by the known facts relating to the situation;
- understands that there are certain acceptable ways of proceeding:

| Moral components | Hypothetical response | Actual student response |
|---|---|---|
| Phil (HC) | 'The man is a human being with distinctive needs.' | 'Even if he was an illegal immigrant he was still a human being and he needs help.' |
| Phil (CC) | 'An illegal immigrant must be subject to the rule of law and be deported.' | 'He is an illegal immigrant and he should not give him money because he was not, by law, supposed to be in the USA and he would be helping somebody who should not be in the country.' |
| Phil (RSF) (DO) and (PO) | 'It is Delaney's duty, being rich, to help meet the needs of the poor Mexican.' | 'The man was in trouble and in need. Delaney should give him money because he could be alone and have no friends and family.' |
| Emp (HC) | 'I would feel frightened getting out of a car on a lonely road.' | 'Delaney should not give the Mexican money because he might have deliberately run into the car so that Delaney would feel sorry for him.' |
| Emp (C) and (UC) | 'I would feel sorry for the poor Mexican especially as he would be suffering great pain from his injuries.' | 'I am annoyed that Delaney thinks first about his car and not the injured Mexican.' |
| Gig (KF) and (KS) | 'There is blood coming from his mouth and therefore he requires immediate medical treatment.'<br><br>'If the accident was reported to the police he would certainly get deported.' | 'He could pay the man's medical bills for him.'<br><br>'Delaney should not give the Mexican money because he might be an illegal immigrant and Delaney could get into trouble for it.'<br><br>'He needs to see a doctor because he has blood in his mouth.' |
| Gig (VC) and (NVC) | 'I think Delaney should get out of the car, apologise to the poor Mexican for running him down, show sympathy for his position and arrange for an ambulance.' | |
| Krat | 'Delaney ought to help the poor, injured Mexican in his situation. It is his moral duty.' | 'I think Delaney ought to give him money for food, for his family and for proper shelter.' |

**Figure 9:** 'Pilgrim at Topanga Creek' and its moral components according to one group of year 8 students.

- possesses an awareness of the moral dimension to the situation and a willingness to make and defend a moral judgement.

The extent to which a student's response does not demonstrate or only partially demonstrate these attributes would be, for Wilson, a measure of the quality of his or her moral thinking.

By using a moral dilemma, such as 'Pilgrim at Topanga Creek', as the focus of geography class work, the teacher will be making explicit the moral dimension to the topic of migration, raising the awareness of the students to the arguments which can be advanced to support or reject the moral decision made by Delaney, and inviting the students to revise their views in the light of the whole-class discussion and further information from Candido's story. This process will help enhance the quality of student's moral reasoning, evidenced in this case by the extent to which this reasoning embodies these moral attributes.

Cartoon: Dave Howarth.

# 6: Creating a moral dilemma

This chapter describes how to construct a moral dilemma for use in the geography classroom in eight steps.

## Step 1: Identify the issue

Identify an issue with a geographical or environmental dimension. Good sources are the local and national newspapers, geographical journals/magazines, radio and television. Newspapers often offer topical issues which relate to themes you may be covering in class. For example, in an article in *The Guardian* (4 April 2000), John Vidal focuses on the dumping of food after it has reached its sell-by date, when there are thousands of homeless and destitute people suffering from hunger (Figure 10). According to the criteria set out in Chapter 3 (pages 21-22), this is clearly a moral issue.

**Figure 10:** An example source for a moral dilemma.

> # Scandal of the food Britain throws away
>
> *by John Vidal*
>
> Britain is throwing away up to 500,000 tonnes of edible food each year – a scale of waste that has shocked charities working with the country's 13 million people in poverty.
>
> Only about 3000 tonnes is given to charities and local authorities for redistribution, and all the charities licensed by the government to collect and store surplus food say that they could distribute much more, were it made available.

## Step 2: Relate it to your scheme of work

Locate the issue within the content frameworks of your school or department's schemes of work for the appropriate key stage or the 16 to 19 age range. In essence, integrate the dilemma within the context of your teaching programme.

# Step 3: Locate other resources

Research and assemble additional background resources for constructing a considered, informed and believable moral dilemma. For example, in relation to the issue of food dumping, Sustain (the alliance for better food and farming) produces a range of publications, including: *Food Poverty: What are the policy options* and *Too Much and Too Little? Debates on surplus food redistribution* (Hawkes and Webster, 2000; Sustain, 2000); likewise, Crisis (an organisation concerned with the needs of the destitute and homeless) has produced a pack for schools entitled *Changing Lives* which includes worksheets for teachers (Crisis, 1999).

# Step 4: The key moral decision

Identify the key moral decision which has to be made and list the arguments for and against it. In this case it is 'whether to distribute redundant food to homeless and destitute people or not' and some of the moral requirements linked to it are shown below.

*For food distribution:*
- the requirement to feed people in need while at the same time saving waste food
- the requirement to put moral pressure on companies and governments to reduce waste
- the need to raise the morale of employees in shops and restaurants
- the need to raise the moral profile of companies as socially responsible organisations
- the need to raise the awareness of young people in school to the issue
- the need to improve the environment by minimising landfill requirements
- the need to reduce the food industry's disposal and storage costs
- it does not require a large bureaucracy
- it can promote local community schemes, e.g. community cafés, co-operatives, demonstrating concern and solidarity for others
- it encourages the homeless to visit day care centres where other services are available

*Against food redistribution:*
- it may perpetuate poverty rather than promote long-term solutions
- it may perpetuate a state of dependency in the homeless and destitute on food distribution schemes
- it may undermine the development of self-help schemes and community-based food initiatives
- it may, because of its reliance on private charitable schemes, unwittingly undermine the safety net of public welfare programmes and allow governments to save money on welfare programmes. In other words it may depoliticise the issue of hunger
- it deflects from the issue of waste generation in the food production system
- it undermines the need for the reform of the social security system and the concept of a food rights and entitlement;

- it reflects and perpetuates inequalities in society
- it entrenches a 'two tier' food system with the 'good' food going to those with money (the first tier) and the 'leftovers' going to the poor (the second tier).

# Step 5: Create biographical details

Identify the key person(s) in the dilemma, i.e. the one(s) forced to make a moral decision. Extract or write brief biographical details for each one. At this stage, it is a good idea to start drafting your narrative. You will, however, need to structure it so that the key moral agent(s) is/are forced to take a moral decision in the context of an unfolding narrative. You must also ensure that students can make the necessary links between different areas of the narrative.

# Step 6: Provide contextual detail

As you are drafting the narrative, be sure to infuse it with sufficient contextual detail on the moral requirements linked to the alternative courses of action or decisions. You can supplement the narrative with additional material, for example, graphs, maps, brochure extracts, photographs.

# Step 7: Revise your narrative

Your narrative will need to be revised before it is ready for use. As you read and revise it, beware of constructing the narrative:

- to give more weight to the moral requirement(s) linked to one decision or course of action, thereby weakening the strength of the dilemma,
- with over-contrived situations or circumstances in the context of the dilemma – this will give it a sense of unreality and it must be believable to work in the geography classroom, and
- with insufficient contextual details – this may include detail contained within the context of the narrative.

# Step 8: Use the dilemma

Integrate the use of the moral dilemma into teaching the geographical topic in order to highlight the moral dimension to the issue.

Photo: Michael McPartland.

# 7: Conclusion

This book has argued that the moral dimension to many of the topics and themes taught in geography ought to be made more explicit and that moral dilemmas are a powerful device for meeting this objective. It has demonstrated that they are rendered more meaningful when they are embedded in a narrative structure, and that by focusing on them the moral awareness and moral thinking of young people might be enhanced and transferred into other contexts.

The use of moral dilemmas embedded within a narrative context serves to reconcile two of the prevailing paradigms in moral education: the cognitive development (Kohlberg, 1981) and socio-cultural development paradigms (Gilligan, 1982). Kohlberg's central assertion is that the use of moral dilemmas to promote moral reasoning in the classroom may lead the student to a higher stage of moral awareness. However, the moral dilemmas used within this paradigm often lack contextual detail, diluting the complexity of the moral choices to be made. Narrative provides that level of contextual detail. In contrast, the socio-cultural paradigm of moral development argues for the importance of the particular circumstances surrounding any moral decision, the inevitability of an emotional dimension intruding on that decision and the need to ensure that, in taking the moral decision, the prime objective should be maintenance of the relationships between those involved in the decision. This paradigm of moral development elevates the ethics of care rather than, as in the cognitive development paradigm, the ethics of justice. In the 'Pilgrim at Topanga Creek' moral dilemma those that align themselves with the cognitive development paradigm would be concerned with the legal position (e.g. Candido is an illegal immigrant, therefore justice demands that he be deported); while those that subscribe to the socio-cultural paradigm would be concerned with Candido's family circumstances (e.g. Candido's wife is pregnant and needs care). Since all narrative is located in a social-cultural context and reflects the particularities of that context it, too, can be used to satisfy the demands of the socio-cultural paradigm of moral development. In a sense narrative serves to reconcile the two dominant paradigms.

Moral dilemmas can be used in geography classrooms in association with a range of teaching strategies: role play/drama, whole-class debates, decision-making exercises, group presentations, paired work, writing activities for specified audiences, imaginative extension work (e.g. constructing a linked sequence of moral dilemmas) and enquiry-based approaches (e.g. searching for new evidence). Moreover, the use of information and communications technology resources can be incorporated into these strategies (e.g. searching for new evidence on the internet).

Finally, the QCA's *Education for Citizenship and the Teaching of Democracy in Schools* (1998), lists the following range of elements under these three headings which, it argues, must be addressed by the end of compulsory schooling:

*1. Knowledge and understanding*
- Topical and contemporary issues at a range of scales

- The nature of diversity, dissent and social conflict

- The nature of social, moral and political challenges faced by individuals

- The economic system as it relates to individuals and communities

- The rights and responsibilities of citizens

- Sustainable development and environmental issues

*2. Skills and aptitudes*
- The ability to make a reasoned argument both verbally and in writing

- The ability to co-operate and work effectively with others

- The ability to tolerate other viewpoints

- A critical approach to evidence and an ability to look for fresh evidence

- The ability to identify, respond to and influence social, moral and political situations

*3. Values and dispositions*
- Belief in human dignity and equality

- Judging and acting by a moral code

- Courage to defend a point of view

- Willingness to be open to changing one's opinions and attitudes in the light of discussion and evidence

- Concern for human rights

Using moral dilemmas, such as, for example, 'Pilgrim at Topanga Creek' (pages 28-29 and 33), in the geography classroom will ensure that most if not all of these elements are clearly addressed, demonstrating the contribution which geography can make to citizenship education.

# Bibliography

Blatt, M.M. and Kohlberg, L. (1975) 'The effects of classroom moral discussion upon children's level of moral judgement', *Journal of Moral Education*, 4, 2, pp. 129-61.

Bottery, M. (1990) *The Morality of the School: Theory and practice of values education*. London: Cassell.

Boyle, C.T. (1995) *The Tortilla Curtain*. London: Bloomsbury.

Bruner, J. (1986) *Actual Minds and Possible Worlds*. Cambridge: Harvard University Press.

Crisis (1999) *Changing Lives* (video). London: Crisis.

DeHaan, R., Hanford, R., Kinlaw, K., Philler, D. and Snarey, J. (1997) 'Promoting ethical reasoning, affect and behaviour among high school students: an evaluation of three teaching strategies', *Journal of Moral Education*, 26, 1, p. 520.

DfEE (1999a) *National Curriculum: Handbook for secondary teachers in England (KS3&4)*. London: DfEE.

DfEE (1999b) *National Curriculum: Handbook for primary teachers in England (KS1&2)*. London: DfEE.

Gilligan, C. (1982) *In a Different Voice*. Cambridge MA: Harvard University Press.

Gowans, C.W. (1987) *Moral Dilemmas*. New York: Oxford University Press.

Hardy, B. (1968) *The Collected Essays of Barbara Hardy, Volume 1*. Brighton: Harvester.

Hawkes, C. and Webster, J. (2000) *Too Much and Too Little? Debates on surplus food redistribution*. London: Sustain.

Kohlberg, L. (1981) *Essays in Moral Development, Volume 1: The philosophy of moral development*. San Francisco: Harper & Row.

Labov, W. (1972) *Language in the Inner City*. Philadelphia: University of Pennsylvania Press.

Leat, D. and Nichols, A. (1999) *Theory into Practice: Mysteries make you think*. Sheffield: Geographical Association.

Lickona, T. (1991) *Educating for Character*. New York: Bantam.

Longacre, E. (1982) *The Anatomy of Speech Notions*. Lisse: Peter de Ridder.

MacIntyre, A. (1981) *After Virtue: A study in moral theory*. Indiana: Notre Dame Press.

Mason, P. (1995) *Tourism: Environment and development perspectives*. Godalming: World Wide Fund for Nature.

Maye, B. (1984) 'Developing valuing and decision making skills in the geography classroom' in Fien, J., Gerber, R. and Wilson, P. (eds) *The Geography Teacher's Guide to the Classroom*. Melbourne: Macmillan, pp. 29-43.

National Curriculum Council (NCC) (1993) *Spiritual and Moral Development*. York: NCC.

Ofsted (1994) *Spiritual, Moral, Social and Cultural Development: An Ofsted discussion paper.* London: Ofsted.

Pring, R. (1984) *Personal and Social Education in the Curriculum.* London: Hodder & Stoughton.

Proctor, J.D. and Smith, D.M. (eds) (1999) *Geography and Ethics: Journeys in a moral terrain.* New York: Routledge.

QCA (1997) *The Promotion of Pupils' Spiritual, Moral, Social and Cultural Development.* London: QCA.

QCA (1998) *Education for Citizenship and the Teaching of Democracy in Schools.* London: QCA.

QCA (1999) *Preparing Young People for Adult Life.* London: QCA.

Sinnot-Armstrong, W. (1988) *Moral Dilemmas.* Oxford: Oxford University Press.

Slater, F. (1982) *Learning Through Geography.* Oxford: Heinemann.

Slater, F. (1996) 'Values: towards mapping their locations in geographical education' in Kent, W.A., Lambert, D.M.L., Naish, M. and Slater, F. (eds) *Geography in Education: Viewpoints on teaching and learning.* Cambridge: Cambridge University Press, pp. 200-30.

Sustain (2000) *Food Poverty: What are the policy options?* (A discussion document drawing together over 70 referenced proposals for tackling food poverty.) London: Sustain.

Tappan, M.B. (1990) 'Hermeneutics and moral development: interpreting narrative representations of moral experience', *Development Review*, 10, pp. 239-65.

Waugh, D. and Bushell, T. (1992) *Connections.* Cheltenham: Stanley Thornes.

White, H. (1981) 'The value of narrativity in the representation of reality' in Mitchell, W. (ed) *On Narrative.* Chicago: University of Chicago Press, pp. 13-14.

Wilson, J. (1990) *A New Introduction to Moral Education.* London: Cassell.

Wright, D. (2000) *Theory into Practice: Maps with latitude.* Sheffield: Geographical Association.

## Contacts

Crisis, Challenger House, 42 Adler Street, London E1 1EE. Tel: 020 7655 8300.
Sustain, 94 White Lion Street, London N1 9PF. Tel: 020 7837 1228.